Alex Morris

The Role of Certification in Promoting Sustainable Golf Courses

AF167271

Alex Morris

The Role of Certification in Promoting Sustainable Golf Courses

LAP LAMBERT Academic Publishing

Impressum / Imprint

Bibliografische Information der Deutschen Nationalbibliothek: Die Deutsche Nationalbibliothek verzeichnet diese Publikation in der Deutschen Nationalbibliografie; detaillierte bibliografische Daten sind im Internet über http://dnb.d-nb.de abrufbar.
Alle in diesem Buch genannten Marken und Produktnamen unterliegen warenzeichen-, marken- oder patentrechtlichem Schutz bzw. sind Warenzeichen oder eingetragene Warenzeichen der jeweiligen Inhaber. Die Wiedergabe von Marken, Produktnamen, Gebrauchsnamen, Handelsnamen, Warenbezeichnungen u.s.w. in diesem Werk berechtigt auch ohne besondere Kennzeichnung nicht zu der Annahme, dass solche Namen im Sinne der Warenzeichen- und Markenschutzgesetzgebung als frei zu betrachten wären und daher von jedermann benutzt werden dürften.

Bibliographic information published by the Deutsche Nationalbibliothek: The Deutsche Nationalbibliothek lists this publication in the Deutsche Nationalbibliografie; detailed bibliographic data are available in the Internet at http://dnb.d-nb.de.
Any brand names and product names mentioned in this book are subject to trademark, brand or patent protection and are trademarks or registered trademarks of their respective holders. The use of brand names, product names, common names, trade names, product descriptions etc. even without a particular marking in this works is in no way to be construed to mean that such names may be regarded as unrestricted in respect of trademark and brand protection legislation and could thus be used by anyone.

Coverbild / Cover image: www.ingimage.com

Verlag / Publisher:
LAP LAMBERT Academic Publishing
ist ein Imprint der / is a trademark of
OmniScriptum GmbH & Co. KG
Heinrich-Böcking-Str. 6-8, 66121 Saarbrücken, Deutschland / Germany
Email: info@lap-publishing.com

Herstellung: siehe letzte Seite /
Printed at: see last page
ISBN: 978-3-659-49315-7

Abstract

   With current sustainability goals, changing the trends of golf course design and management is an important issue. Golf courses are a good focus because they are important socially, economically, and environmentally. This paper investigates Audubon International golf course conservation certification's role in assisting golf courses ecological well-being; and how this influences sustainability trends. A research review shows Audubon International is up to date on resource conservation practices. Consumers are important in reinforcing the trend of courses perusing certification. Consumer data is still needed to indicate if environmentally friendly design is highly demanded. If so, these courses will not only better conserve resources, but also gain a competitive advantage over other golf facilities and become more economically viable. Golf courses are already municipal revenue and environmental assets, conservation certification can further help these facilities achieve sustainability. Included is a case study of two courses within the same municipality that show higher revenue at the certified course. This reflects some support from the Jefferson County Open Space Fund for the courses outstanding wildlife habitat, but also likely the consumer's enjoyment of the course. However, it remains unknown if this support is in response to the conservation certification specifically.

Table of Contents

Introduction

Topic Introduction

There are a variety of sports, but golf is arguably the only sport with advantageous ecological attributes. Sustainable development of golf courses requires analysis of the climate, management, and the functionality of the course design within a city. I will explain how Audubon certification attempts to optimize courses economically, while also creating a more ecologically and socially valuable land use. Economic criteria are no longer sufficient; we also need to address issues of equity, environmental concerns, and the course social benefits. The stewardship of land, water, and biodiversity is the future of course development in order to preserve these resources as they slowly are depleted. After creating strategies for improving design and managing resources, their implementation needs to be incentivized. One incentive is lowering the expenses of the course through resource use efficiency. Another incentive is to create a course that attracts ecological consumers. What follows is that environmental decisions from ecologically responsible players and courses also benefit society.

Golf courses are an environmental asset to city municipalities. State and Federal governments have invested in large-scale parks that preserve habitat and important landscapes. County governments work with open space areas that are slightly less substantial, but still serve ends that are related to both environmental conservation, and outdoor recreation. Golf courses are ideal for cities because they not only provide a functioning habitat, but also a recreation activity that appropriately creates revenue at the smaller scale. Golf course development is desired by voters, and has entertainment value as well as significant revenue. Also as a public good, there is excludability with green fee restricted accesses, and this does not allow for a free rider problem. The concern is that turf management practices that involve chemical and water are environmentally damaging.

Purpose of Research

I will explain how Audubon International golf course certification is an incentive for sustainable golf course management. However, consumers create development trends and their awareness and behavior regarding certification will determine certification success. Courses that consider the environment will more easily receive land permits for construction. Certificates have four different rankings once awarded, and are based on resource criteria, optimal local design, and hedging environmental externalities. These same criteria are also important for courses considering certification.

The ideal process relating to golf course sustainability needs to be identified. The ideality involves improving resource conservation to a point considered sustainable, while still achieving customer satisfaction. There are three primary areas that require attention: local climate and species, design functionality, and adaptable management. Every area is unique making the process more difficult. Like many solutions, no one single method will succeed everywhere. The Audubon certification is devoted to an individual analysis of all the course locations to account for the variable conditions of each course. This provides better confidence to consumers paying for a certified course. As the leaders of course certification, an analysis of an Audubon endorsed course can illustrate their benefits. The goal is to prevent 'green washing' the process of marketing environmental practices, which has no actual benefit. A course cannot persist on environmental policies alone, so in addition to environmental data the course economic history and consumer preferences will contribute to the sustainability assessment of the course.

The Audubon International conservation certifications are the primary mechanism for creating environmental golf courses. My goal is to analyze certification effectiveness in order to make sure environmentally progressive courses are available. The Audubon International procedure is practical because they do not subscribe to a singular solution for different regional problems. They determine what is appropriate on a case-by-case basis that treats each area differently from the last, using indicators like climate, flora, fauna, and resource availability. Further, the evaluation matches the triple bottom line for sustainability, considering social and economic value in addition to the ecology.

The knowledge gained may build confidence and reliability to assist players with environmental preferences, and will allow for a future progressive environmental course strategy. The social preference of recreational golfing as well as the economic viability of thousands of constructed courses is readily observable. Task one is to identify if certified courses work more harmoniously with the surrounding habitat. The second is to address whether consumers prefer more ecological courses. The implication of these together will identify the potential of higher revenue corresponding to increases in demand. This also requires marketing of the certification to build awareness, as well as consumer confidence in the certification agency.

Further, through managing resources more effectively the cost savings observed in certified courses can possibly incentivize modifications that improve environmentally damaging courses. The savings needs to be tangible, because some managers need proof of long-term savings in order to pursue redevelopment, which requires a high initial investment. The study and evaluation of the differences between two similar courses has the potential to show these benefits in the certified course that can inspire improvements in others. The courses I have chosen are within close proximity, with the same greens fees, and managed through the same City municipality; however one has the bronze certification and is one of the two highest Audubon certified courses in the state of Colorado, and the other has no certification.

The intention is to widen the environmental debate concerning golf courses. Various benefits will be contrasted with other perceived costs. The evaluation will determine if the progress made concerning course environmentalism has the potential to achieve sustainability. Peer-reviewed research is essential to illustrate modern improvements in courses such as water use efficiency, design, land valuation, management, and runoff. This analysis will require a qualitative review to contrast how damaging or beneficial these course practices are on the environment. As well, those preferred within Audubon certification will be incorporated to better determine the effectiveness of this institution.

Social research will need to assess golfs benefits and costs relating to health, as well as ideals and values, and equity issues. For example, while golf can be a healthy form of exercise, chemical applications can negatively affect one's health while walking and interacting in the environment. In addition, ideals and values that stem from golf such as stewardship, integrity, practice, and self-discipline, potentially carry over with an individual and benefit society. However, there are equity issues between those who directly benefits from golf, and others who negatively associate the game as an elitist activity. What partially offsets this debate is the money raised through charity golf tournaments held nationwide that are intended to contribute to the underprivileged.

Background

Golf was a pioneer sport when originally developed in Scotland and Ireland. Links courses, a style of construction on coastlines, work with the land, and provide entertainment. Their success relates to an effective use of the land that is otherwise too unstable for structural development and the water too rich in saline for agriculture. Grass has salt tolerance and some, like the Spartina genus, compose salt marshes. Grazing maintained the turf for play as well as provided the communities with healthy livestock. The original form of golf was socially and environmentally useful. However the modern golf course does not achieve the same level of functionality. Modern luxuries have moved courses away from the sustainable historical model. Technology requires longer courses, this influences energy used in mowing and chemical applications.

Currently there are numerous courses globally, and these can be used as ecologically friendly recreation areas, or unnecessary resource expenditures. With the quantity of global courses, redevelopment is encouraged rather than constructing new courses with modern design knowledge.

Even courses built in the 1990's were as ecologically irresponsible as other golf construction boom

periods in the 1920's and 1960's (Hueber 2010). Courses like these need ecological adaptions.

> *"The golf courses were not environmentally sensitive, economically viable or socially responsible.*
> *Too many golf courses were built and more often than not they were built in the wrong places.*
> *Too much money was spent on building the courses and the maintenance of these golf courses is*
> *also expensive, so now they are too expensive to play. Additionally, many of these golf courses*
> *were too difficult to play for the average golfers"* (Hueber 2010).

The above factors influence golf facility closures in the 21st century. Singer and Richardson (2009) note

that the supply of courses greatly increased in the 15-20 years preceding the loss of golf profitability

observed in the 21st century.

Conservation considerations allow for golf course specialists to consult on a property in order to

identify management goals, plan to use resourcesbetter, and offer reassurance to environmentally

minded players. The period of 2000-2003 noted a decrease in roughly three-thousand "golf rounds per

18-hole equivalent," and this drop has remained constant between 2003 and 2008 (Hueber 2010). With

golfing desirability decreasing, an environmentally sound course is more likely to have support from the

community and players. Modifications are required for the future of golf course sustainability, as Jim

Hyler, the President of the US Golf Association indicates, "In my opinion, many of the standards by

which we construct and maintain our courses have become, quite simply, unsustainable" (Grooms, 96).

The Audubon certification program provides the correct institutional diversity to promote more

sustainable trends in golf course development and management. The goal is to determine whether a

certificate process can switch the trends of course development in a sustainable direction.

Data Collection

The main focus of this research is the semi-arid climate of Colorado. Problems of water scarcity

are the front line of attacks on golf course land uses. A peer-reviewed data collection will address this

primary concern as well as: pollution, land valuation, and course design. Original data collection will

juxtapose two municipal golf courses. Course A, is an Audubon bronze certification with other variables of management and customer base relatively equal to Course B, which has no external certification. The analysis will scrutinize the success of this certification along the triple bottom line. The economic history will show not only revenue but also course budget in order to determine if a market exists for environmentally progressive golf course designs. The budget indicates how much outside inputs are required to operate the course. A more environmentally friendly course should indicate fewer inputs, and thus a lower budget.

Water Use Planning

Waste Water Reclamation

Water is the primary resource under investigation for improving golf course revenue and environmental impact. Wastewater reclamation is positive environmentally, because the process counters negative runoff, while recycling nutrients and purifying grey water. This conservation strategy is best for municipal golf courses and developers that maximize the potential of the landscape in the golf course design. Storm runoff requires retention ponds designed to maximize storage. The storage pond can capture housing development effluent, while also reusing excessive chemical applications. Reusing wastewater is an investment when designed to fit the location of the course. Both the water conservation and turf chemical recycling are cost effective, while grey water use and ground water percolation are social services. "Colorado courses tested for long term effects of grey water use did show a 187% salinity increase from the regular surface irrigation" (Qian et al. 717, 2005). Qian and Mecham (2005) also found high sodium and beryllium accumulation; the side effect is potentially useful when courses are managed with inherently salt tolerant grasses. Salinity is a useful herbicide when used correctly. Irrigation of a 100% seawater solution has grasses recover whereas the weeds do not (Couillard et al. 1998). The correct grasses in combination with periodic leaching to decrease salt accumulation, percolation drainage, also regular soil and plant monitoring (Qian et al. 2005), can use salinity beneficially as a weed deterrent. Finally, Qian et al. (2005) recommend a sulfur burner to mitigate salinity and pH problems. A modification used at course A, which has the conservation certification.

Reclaimed water requires diligent understanding of several key variables. The climate is a specific constraint. Semi-arid and arid courses require artificial irrigation and can benefit the most from recycling water. The specific quantities of water, chemicals and seeds in conjunction with the seasonality of the course, directly correlate to the total cost of course operations. The processes of analyzing the variables are difficult, but when operated ideally, costs are minimized. Golf courses are very complex systems and long-term studies are expensive. King et al. (1999) tested parameters of the necessary variables, with sufficient data to justifiably set parameters in a computer model. The goals, as well as the recommendation, seek to improve models that evaluate the specific levels and times of applying fertilizers in conjunction with reclaimed water (King et al., 1999). This group acknowledges the importance of better understanding best management practices for turf grass systems as a way to improve efficiency and minimize costs. King et al. (1999) present just one strategy to reduce management practices for turf grass systems. King and Balogh clearly state that pesticide in combination with irrigation has not been studied fully; as well the study is limited in reviewing wetting agents or growth regulators in combination with salinity mitigation to maximize the potential for reclaimed water use. However, in terms of over seeding, fertilizer, and water use, this study is a valuable step in the right direction. The ultimate goal is to combine the seasonality and application schedule variables to improve models, with further data, and more management scenarios. Through understanding how these variables interact in a turf grass system, managers can make better economic and environmental decisions.

Managing chemical applications for water use efficiency

Hydrophobic soils repel water not only contributing to dead turf, but also excessive runoff. To conserve water many researchers are analyzing causal factors as well as solutions for hydrophobic soils specific to golf courses. Relevant research regarding managing hydrophobic soils began in the 1990's and still the most common management strategies are not sufficient. Wetting agents are the primary tool for reducing the localized dry spots that occur on hydrophobic soils. Some research suggests that wetting agents do not deal with the cause of hydrophobic soils. The article "Microbial Derived Water Repellency In Golf Course Soil" (Hallet et al., 2001) outlines the study of the cause of hydrophobic soils through monitoring soil fauna under controlled fertilizer regimes of 25 kg N/ha with specific carbon to nitrogen ratios. The results are insightful,

> "Severe levels of water repellency only developed for the highest carbon substrate addition (18mg/mg soil) for all soil except when treated with a fungal biocide. Soil treated with a bacterial biocide developed higher levels of water repellency than an untreated soil, suggesting that microbial competition can suppress the development of water repellency... [and] suggests that fungi are the component of the microbial community that causes water repellency" (Hallett et al, 2001, 518).

This article progressed the understanding of hydrophobic soils at a time when research focused on wetting agent solutions was just beginning. Wetting agents still are a common practice "87% of more than 600 superintendents use wetting agents as regular maintenance" (Karnok et al., 2004) even though wetting agent use is potentially unnecessary or overvalued.

Karnok and Tucker are prominent wetting agent researchers, publishing 5 articles together in Golf Course Management and Karnok having two others as well, but the majority of their work was done after fungus was identified as the cause of hydrophobic soils. Is their advocacy of wetting agents unwarranted? Karnok's early investigation of localized dry spots was published in 1995. Useful findings regarding hydrophobic soils include the fact that it occurs on soil with a high sand content, is not greater than three inches deep, and occurs 6 to 18 months after construction regardless of fertilizer and pesticide use (Karnok et al 1995). Although not identifying the fungal cause, these characteristics are useful for managers. The earliest variables at fault that require monitoring are; irrigation practices, soil pH, particle size, and species of turf grass; these are incorrect but a good starting point for study. In 2002, Tucker and Karnok released a two part series in golf course management. In part one, their addition was using wetting agents as a cure, although the year prior, fungi were identified as the cause. Their two most useful contributions are that excessive wetting agent applications are a waste and unnecessary cost; as well, in part two they surveyed 300 golf course superintendents regarding wetting agents in order to address the frequently asked questions. In 2003, Tucker and Karnok recognize the importance of extensive rooting of turf grass to reduce stress, this is where wetting agents are useful, but growth regulators also contribute to root growth. Since this is not countering the source of the hydrophobicity there is still some controversy about whether these chemicals are the best management practice. However, fungi are not the only turf grass stress. Cultural practices for golf courses stress golf course greens. Manager's cut grass short, at an eighth of an inch or lowerto maximize speed ball. This practice limits photosynthesis, dries greens to critical levels, and reduces aeration, all restricting root growth (Karnok et al 2003). In 2004, Karnok et al. investigated the national use of wetting agents and reported that 87% of managers use them regularly and 11% on certain occasions together comprising 98%; 42% primarily use them for localized dry spots, 32% for managing water, 9% for improving pesticide movement into the soil, and 11% to improve drainage. The 2006 publication was a review and list of factors superintendents must consider when choosing one of more than 200 wetting agents that range in cost from $25-$100 per gallon. With various of brands and multiple wetting agent uses, an acreage of maximum use regulation, abide in society's best interest. The concern is that Karnok et al. (2006) recommend supporting research for decisions as step seven, but not all agents are tested and effects remain largely unknown. In 2008, Karnok and Tucker wrote their final defense of using wetting agents in Golf Course Management, stating 50% of the water was required when using a wetting agent to reach 15% moisture content of the soil. Since hydrophobic soils are such a common issue, wetting agents seem useful as a trade-off for water conservation and efficiency. However, "greater than 80% of the wetting agents on the market are not all tested this way" (Karnok et al 2008). Policy is necessary to require greater preliminary investigations before wetting agents go into production and widespread use.

Growth regulators, as mentioned previously, are used to improve turf system functioning and water conservation. The goal for chemical application research is to clarify best management practices because turf systems are complex and conditions vary among courses. Growth regulators are just a part of the solution, and quantity to formulation ratios is not guaranteed. Most important for golf courses in general is that growth regulators increase the roots density and toughen the turf in extreme heat and play (Shepard et al., 2000). In terms of water conservation, growth regulators can help courses a third way, because water requirements are reduced roughly seven to 26 percent with equivalent quality, while transpiration is reduced because leaf surface area is reduced (Shepard et al., 2000). The combination of wetting agents and growth regulators influences golf course efficiency and uncertainty still remains. One highlight of the Shepard et al. (2000) article is a correlation between positive fungicide efficacy and growth regulator use, which in combination could more easily combat hydrophobic soils, though more research is required. In an earlier publication, Fry and Hongfei (1998), corroborate the inadvertently reduced water requirement because fast growing grasses have higher evapotranspiration. This is an important accommodation to semi-arid golf courses that allow for acceptable water use. Further, growth regulators reduce clippings, fewer clippings more easily decomposes on course, and this ability to mulch grass is a recycling technique. The collaborative research required to identify wetting agent, growth regulator, and fungicide regimes has the potential for higher efficiency as management regimes adapt.

Better water management

Attempts have already been made relating to best management practices to construct strategies for water conservation. Carrow et al. (2005) outline their target areas for managers pursuing full scope best management practices. The process asserts three procedures. First, the site assessment and planning involves the use of native grass species, reduced irrigated turf, improved irrigation design, and weather/seasonality-based scheduling (Carrow et al., 2005). Second, the identification, evaluation, and selection of water conservation strategies are currently staged for analyzing growth regulators and wetting agents (Carrow et al., 2005). The third procedure is to assess the benefits and cost to all stakeholders before implementation (Carrow et al., 2005). What is fundamentally missing is raw data on the various chemical products, and even less analysis of how these agents work in combination. Research is fundamental to discovering the best management of any system. Carrow et al. (2005) agree that regulations and mandates for golf courses are essential for long-term sustainability, requiring more rigid chemical evaluation. Research into drought resistant grasses is less necessary, although done to a similar extent, but still valuable to managing course water use. The Mintenko and Smith (1999) article "Evaluation of Native Grasses for Low-Maintenance Turf" focuses specifically on four drought resistant species due to the long-term water threat to the turf industry. Most notably is the St. Augustine grass, remarkably drought tolerant, surviving 158 days of successive drought stress, with the total success of leaf firing after 158 days compared to a thirty-day recovery (Mintenko et al., 1999). Research is a key to efficient management, and the interaction of all these variables is more easily examined in a computer model due to system complexity.

Hydraulic lift is a phenomena not currently researched in relation to golf courses, but important for vegetative systems. Hydraulic lift happens when a deeply rooted plant extracts deeper water resources and excretes them near the surface, introducing more available resources in the system. The planting of trees on golf courses will have an effect on water use and need consideration in planning and management. Grasses do use hydraulically lifted water from trees, this process strengthens root systems, but the usefulness is easily overwhelmed in semi-arid or arid regions (Ludwig et al., 2004). Course A utilizes native grassland surrounding the turf, while small quantities of trees are found in the riparian areas; this design effort likely contributes to their conservation standing. The issue is how hydraulic lift from trees is studied; most researchers including Ludwig et al. (2004) are researching the effect in natural ecosystems. The research could prove more useful relating to managed systems, especially when applied to the expansive acreage of turf grass globally. The Ludwig et al. (2004) study in particular focused on east African Acacia trees and the surrounding savannah grasses; however this semi-arid steppe region does have some climate similarities to Colorado. Alternatively to conserving water through the use of hydraulically lifted water, grasses also showed deeper rooting to access water reserves when in direct competition from trees. If some of these principles apply in turf systems, then this is another strategy for consideration to conserve course water, and in turn, course revenue. The environmentally positive attribute of water conservation will likely compensate the course for lost aesthetic value. However, better marketing data is needed to identify golfer's choices. Like all decision-making processes, trade-offs are to be considered and tree water consumption, which has higher evapotranspiration than grasses; must be weighed with the benefits of shade and reduce turf water use.

The Applewood Golf Course in Golden, Colorado is a role model for a semi-arid course. This course changed to, "only 11 ha of 'pampered' turf that has drastically reduced the use of water and chemicals. The wildlife value of the course has increased, and these reductions in irrigation and chemical inputs have lowered the risk of ground water contamination" (Terman 11, 1997). This course compliments the recreational value of golf with quality habitat, and intelligent resource use. However, this course does not qualify for Audubon certification, or does not value the program highly enough to participate. These modifications are helpful regardless of certification, but the certification allows the course to market their environmental changes and benefit from the certification consultants experience. Mimicking successful modifications from environmental golf courses is helpful in identifying best management practices for other courses, and helps in other management areas, for example, minimizing externalities.

Externalities

Health effects

Turf grass areas are a beneficial resource for recreation only if chemical management of the areas does not negatively affect human health. Pesticide vulnerability is a part of our well-being that no one should overlook. Louise Chawla (2011) has concern, as a member of the Children, Youth and Environments Center for Community Engagement, with an article on "Child Friendly Lawns and Gardens." She recognizes the importance of turf grass resources for kid's health, for exercise and activities, but also states, "the outdoors must be safe from risks that children cannot see and negotiate" (Chawla 2011). Even knowledge of the dangers of the many pesticides does not enable a person to detect the pesticides they interact with or ingest. Several direct health impacts are listed including, "delayed cognitive development, learning disabilities, lower IQ, and attention deficit disorders in exposed children", as well as indirect trauma when loved ones sicken or die from, "cancer, reproductive disorder, immune system disorder, and neurological effects like Parkinson's disease," (Chawla 2011). The widespread use of a multitude of pesticides suggests the need for stringent regulations to minimize people's distress. Chawla (2011) points out the inadequacy of current regulations when considering that product testing is minimal, the consequences of the endocrine disruptions pesticides included are serious and numerous, and even with sufficient evidence of harm, it can take years to remove a toxic product from the market (Chawla, 2011). Commonly, the greens are the most sensitive golf course areas requiring pest assistance. Each course has a practice green and 9, 18, or 36 holes with one green each, and these typically represent 3-5% of the total acreage (Cox, 1991). Using pesticides on these areas will allow the surrounding habitat to absorb and use the excess runoff. For the case of hydrophobic greens, fungicides have the potential to increase water efficiency; however, golfers spend time on every green during a round of golf, implying high player exposure. These chemicals stay active for years and can enter your home on shoes and clothes, and even through windows (Chawla, 2011). The undetectable and active pesticides warrant policy involvement as they directly affect people's rights.

In particular, some argue that golf courses cosmetic standards for turf are unnecessarily high (Cox, 1991) and legislation could limit resource uses. Turf is still playable and regenerative when somewhat aesthetically unappealing. Despite the common use of pesticides and even some identified non-target benefits, these chemicals require strict production and use regulations (Smiley, 1981). Research needs to pursue secondary and even tertiary effects of pesticides, before the real costs and benefits of pesticides are accurately assessed (Smiley, 1981). Minimal use, of thoroughly researched chemicals, is a precautious, but intelligent, strategy for managers until regulations strengthen and further research is available. Finally, Wheeler and Nauright (2006) write "Chatterjee's study published in 1993 that an average of 1500 kg of agrochemicals, some of themknown carcinogens, are applied to golf courses each year and that 90 per cent ofsprayed chemicals end up in the air." This deserves attention, or even legislation, because human health is a primary concern for the policy agenda.

Mitigating Runoff

Aside from pesticide health effects on humans, nutrient run-off can disrupt the balance of surrounding ecosystems. Steams consolidate the effluent carried from courses as well as agriculture systems; polluted from human introduced fertilizers. These historically new changes in the system cause algal blooms, when the primary producers rapidly reproduce in the presence of more available nutrients. This process over-consumes the waters dissolved oxygen in extreme cases, which causes a massive die off of larger fauna, such as fish. The conditions for anoxic water, caused from excessive dissolved nutrients, causing algal blooms is termed 'eutrophication'. The largest algal bloom created at the mouth of the Mississippi river from the United States agricultural land has a noticeable effect on the gulf fishing industry and wildlife. Golf courses currently are not part of the solution to regulating human inputs to natural systems. However, altering golf courses can dramatically influence their contributions. Also, studies like that of King et al.(2001; 2007) monitoring 115 precipitations and run-off events over 5 years at Morris Williams municipal golf course found that "nutrient levels in stream flow exiting the course do not pose immediate health risks." Since healthy limits and accumulation of chemicals are less identifiable, minimal use is a prudent strategy and several methods are in use to control run-off.

Applying chemical to disperse outward from the center of fairways and greens introduces a buffering effect that limits external influence of nutrients. Buffers, in conjunction with other environmentally conscious management tactics, can maximize application utility, and in turn, mitigate loss of nutrients and their associated negative effects. Nutrients are best buffered with three-inch tall grass between turf-grass applications and water bodies. In addition, chemicals are most easily lost when applied to saturated soil; and pesticides with the lowest water solubility and highest absorption coefficients reduce loss (Baird, 1998). The ideal minimum width for a zone buffering water bodies is 30 meters, and when un-mowed or un-maintained the absorption effect is maximized and operating cost minimized (Winter et al., 2005). Further, Terman (1997) identifies how buffers positively affect wildlife management procedures. These basic techniques were reported over ten years ago and are still feasible modifications. All courses will benefit economically and environmentally from incorporating these methods of buffering to reduce loss. Baird (1998) concludes that surface run-off of applied chemicals decreased 2% after the minor improvements.

Buffering does have limitations including an inability to affect subsurface drainage which, in turn, can alter the chemical composition of groundwater. Outflows of chemical applications occur after roughly seven to nine months, however golf courses show nitrates in subsurface drainage at quantities of one-tenth those observed in water leaving crop agriculture (King et al., 2006). Golf courses are a functioning ecological system with less disturbance and nitrogen run-off than crop agriculture which makes it an acceptable land use alternative. The problem on golf courses is that they leach more dissolved reactive phosphorus than in agricultural, based on samples (King et al., 2006). This implies golf course managers need to strictly monitor phosphorus fertilizer use because phosphorus is the limiting nutrient in many wetland systems and increasing it in these areas influences eutrophication.

Wetland retention ponds allow for settling, consumption, and reuse of chemicals like phosphorus, lost in run-off. Or, the digestion of pollutants can restrict the outflow and external impacts of harmful byproducts of golf courses. Mercury is a substantial environmental toxin, and one of the worst pollutants observed from golf course construction and operation.Winter et al. (2005) observing mercury from the use of mercuric fungicides recommended moderating mercury pollution with retention ponds. Also, environmental impact is further reduced when retention ponds are combined with re-use, and waste water irrigation to localize the excesses chemical additions to the course of origin. When trying to alleviate loss "ponds require appropriate management to avoid excess Phosphorus outflow" (Winter et al., 2005). Again, phosphorus disrupts external systems, because it is a common limiting nutrient. Other simple and effective techniques are to use fertilizers as little as possible, and encourage deep rooting trees (Winter et al., 2005). Buffering from trees compliments other proactive measures like monitoring chemical use, and maintaining retention ponds. These adaptions create better managed golf courses.

Golf course redevelopment to accommodate environmental impacts is generally more effective than the construction of new courses. Golf course construction involves many environmental impacts in and of its self. Even designing the course to maximize the environmental potential is not as productive as modifications to the numerous courses in existence. In addition, waste water irrigation retention ponds minimize loss; the ponds capture run-off pollution and help protect the natural balance of surrounding habitats (Winter et al., 2005). Nutrients that are stagnant in a retention pond undergo sedimentation, are consumed, or are available for re-use. Functional retention ponds are an investment opportunity for re-modeling courses, and a must for any modern constructed course.

Ecology

Golf Course Natural Areas

Golf courses provide more wildlife habitat than many other land uses, and the interaction with wildlife is one of the enjoyments for some golfers. Better utilizing and developing natural areas has positive benefits for the course in addition to wildlife. "The use of native grasses to establish attractive, environmentally beneficial, low-maintenance areas is one of the most commonly desired types of

natural areas. Stands of native grasses result in water savings, reduced fuel use and labor, plus improve

aesthetics (Nelson 1, 1997). A course may have the incorrect soil, climate, or management techniques to

install natural areas correctly, so agencies that specialize in conservation, like Audubon International,

can aid courses in assesses the usefulness of different natural areas. Nelson (1997) examines the

establishment and benefits of grasslands, weed control, wildlife, and riparian areas including lakes,

ponds, streams and wetlands, as well as forested natural areas for golf courses. These earlier insights

have enabled proactive change in both course management and design. These strategies are the type

encouraged in conservation, and a consideration for Audubon certification.

Computer models allow for planning course modifications to minimize managed turf, encourage

animal corridors, and use buffers for water bodies. Modern courses should require G.I.S. planning to

best fit the landscape mosaic, because small connected parcels and high quality habitat are ecologically

valuable (Terman, 1997). Computer simulated designs allow for feedback on different construction

options for new courses, as well as course modifications. Although using this type of technology is a

higher cost initially, the programs can trouble shoot design plans prior the courses construction. The

ability to incorporate sensitive habitats around managed turf avoids irreversible changes to wildlife

areas. Further, water use is a feature of course designs, and computer models are useful in evaluating

irrigation, which is an expensive modification. Models increase functionality of the course design based

on characteristics of the local landscape and climate, which can ultimately increase natural areas and

their biodiversity.

Biodiversity

Courses show high bird diversity, because they are sufficient habitat patches, and useful

migration refuge. Golf course conservation certification is applicable to endangered birds. Birds of

conservation concern that commonly do not nest in artificial cavities or do not directly forage on turf grass benefit in nesting success on golf courses, though, in the study, chick survival rates were not significantly different (Leclerc et al. 2005). On average, golf courses have 15 – 22 hectare of unused habitat areas (Leclerc et al,. 2005), and, if designed correctly, habitat connectivity is maximized allowing for populations to persists and stabilize. Because trees buffer water bodies and provide quality bird habitat, location considerations are important. Birch trees are a management concern and should be removed from courses because they excrete germination immobilizers that restrict turf establishment (Nelson, 1997). In addition, addressing other tree impacts can maximize habitat quality and wildlife.

Two important questions Tanner et al. (2004)address are, "Do golf course have higher diversity than farmlands? And, does course diversity increase with age?" Results showed that the birds, and both sensitive insect taxa, had higher richness and abundance on courses (Tanner et al., 2004). Birds have a relationship to tree availability; the larger number of tree species observed on courses explains the habitat niches for bird diversity (Tanner et al., 2004). In relation to diversity with age, the authors noted changes in tree species, and found greater conservation attitudes on new courses. Tree species likely had successional changes over time, while still facilitating the birds. This study specifically concerns the 2,600 United Kingdom golf courses at .7% of the total land cover. Since, golf courses showed more ecological significance than farmland, some golf course land use is desirable. Primarily if courses can support either source, or even sink populations.

Ecological Pest Management

Grasslands out-compete other viable ecotypes through regular wildfires. Open field burning is a very effective means of reducing weed pressure since it destroys weed seed. Burning also reduces disease and insect pests and improves vigor of desirable grasses" (Nelson 2, 1997). However, burning also releases the vegetation's stored carbon. The released carbon is a contributor to global warming, and, therefore, a concern. Using other ecological factors is likely to have smaller effects, but are useful in combination with decreased chemical applications. Courses can lower operating costs by utilizing ecological pest management such as birds to consume insect pests and reclaimed water saline in quantities to act as an herbicide.

Other investigations to reduce costs and chemical management utilize attributes of specific species. For example Gange et al. (2001) studies whether arbuscular mycorrhizal fungi could be used to control the growth of poa annua grass on golf courses. The preliminary results were not strong, but the management applications from similar studies seem promising. The ability to use natural ecological defenses to balance a managed system is highly prized. The process of integrating ecological interactions on golf courses is very complicated, interconnected, and variable because of the many influences from soil, temperature, resource availability, seasonality, and flora and fauna dynamics. In all, ecological management is a complex scenario for golf course superintendents.

Wetlands

In Stockholm, Sweden researchers assessed wetlands on golf courses. An analysis of the area, using GIS, indicates that golf courses contain greater than one-quarter of the permanent fresh water, so wetland fauna can benefit from course water bodies, and wetlands are beneficial regardless of the course location (Colding et al., 2009). Golf course chemicals appear to have little negative affect on the aquatic species, however should not be misused. To further promote wetland conservation on golf courses, ecologists need to cooperate with urban planners, ecosystem managers, and golf course designers. The Audubon program sponsors these interactions and all areas contributed to Course A's bronze conservation certification.

Benefits of Turf Grass

Turf grass has ecological benefits that can offset chemical use or make them unnecessary. Courses generally maintain and use 20-30% of their total acreage. The play is concentrated to putting greens and tee boxes, an average of 3-6 acres per course, and fairways, the other regularly maintained areas take up another 25-30 acres (Beard, 1996). Pesticides, fertilizers, and grooming are most common on greens and tee boxes, because players expect consistency on these areas. Turf grass, discussed before, is a good buffer for chemicals when maintained at three inches and roughly a 30 meter width between water bodies. The grass benefits from application runoff and protects the surroundings from contamination. Minimizing the maintained turf surface allows for lower cost operation in terms of maintenance such as mowing, and the quantity of water and manufactured chemical products. In combination with native grasses, or tree stands, the golf course natural areas serve as high quality wildlife habitat. Nearly 70% of preferred wildlife habitat is woodlands, grasslands, or wetlands (Beard 1996). Humans are another species that value turf grass.

Both physical and mental benefits are observed at golf courses (Beard, 1996). Golf is physically and mentally engaging and is desirable because it is an outdoor activity. The aesthetic green space usually encounters those who pass by as well as players. Turf grass has appropriate benefits that alternative developments do not. Alternative development may be equally required, but golf courses are useful when protecting sensitive habitats that other industries would disrupt. This allows for certain species and soil resources to be maintained for future generations. In addition to erosion control, turf grasses minimize dust and mud problems of un-maintained space. The evapotranspiration of courses can directly dissipate heat and moderate the local temperature. Finally, if used as sound buffers, humans will hear less than if the area was a hard surface (Beard, 1996). In addition, to photosynthesis and respiration, turf grass is a functioning ecological area with high recreational value.

Beard (1996) points out that run-off is minimized except during the most intense storm events, the enhanced soil water infiltration of turf systems allows for ground water recharge. He also indicates urban run-off has more pollutants than courses. Further, reducing insecticides allows for more water holding capacity in the macro-pore spaces created for the system to utilize (Beard, 1996), and soil restoration, improvement, and biodegrading organic chemicals, are also notable turf use observations.

Promising research of native grasses for low maintenance turf is currently underway. With more funding, undomesticated native grasses have the potential to produce turf grass cultivars that thrive with reduced maintenance, and some early concept native cultivars are already available (Mintenko et al., 1999). Native grasses improve course resource uses for water and chemical applications, as well as habitat quality and land use. In addition, Terman (1997) includes support from professional golfers that promote native grass use saying that less managed turf will further push players of all skill levels, and bring back a more traditional Scottish feel. Native grasses are also adapted to yearly cycles of areas and these promote mutually beneficial species interactions.

The phenology of species relates to their annual habitual cycles; for instance, bird migration. As mentioned earlier, chemical applications are better managed when in combination with the natural annual patterns of the turf grass system. Early fertilization can influence the competition dynamic of weeds and grasses. More importantly for Colorado, natural fertilization occurs when geese and other birds migrate annually. Golf courses act as a sanctuary with shelter trees, wetlands, and insects, so, consequently, they consolidate much of the birds' excrement that enables turf fertilization. Birds also are the focus of the Audubon society and are a beneficial wildlife feature on golf courses. Moreover, birds need areas to refuel, and this ecological predation feature enables naturalistic insect control at courses along these migrations.

Land Use

Using degraded lands

Of the safest uses for completed landfills, including ball fields, parks, and playgrounds, golf courses are the most profitable (Aplet & Conn). Problematic consequences of landfill use are subsidence, water pollution, and gas migration. Subsidence is a side effect that severely disrupts development; however golf courses can adapt to changes as long as structures are planned off unstable ground. Water pollution and gas migration are problems at operating on closed landfills despite the above ground land use. Gases are valuable fuel resources that are harvestable if properly ventilated from the ground. Aplet and Conn'sconclusion is these multiuse areas are still insufficient to solve modern land use needs. Golf courses are still reasonable options and can function in these areas. A completed landfill 18-hole golf course in Indiana is an Audubon signature course due to the environmental innovativeness of the combination (Grooms). Terman (1997) agrees, stating, "Degraded lands such as landfills and old mining sites seem ideal for golf courses and naturalistic designs could improve the environmental conditions considerably" (p. 11).

Municipal Recreation and Revenue

The article "Put the Swing Back in Your Game: Eight Ways For Improve Municipal Golf Courses" offers a format to understand how Audubon certification assists the functionality of participating courses. The first tip in Singer and Richardson's (2009) article is "Imitation and Big Budget Mentality (p. 4). Audubon International shows an investment in the environment and their experience of consulting with many diverse courses enables them to recommend imitating other successful course modifications. Tip two, "The Competition" is an Audubon assessment of the level of conservation specific courses should pursue, based on consumer preferences. Tip three "It's Alive!" emphasizes the ecosystem services of turf grass entities, and details specific course deterioration concerns related to irrigation, drainage, and root zones that require attention to maintain the course functionality. Maximizing these components for ecological success is part of Audubon assessments. "Lack of Planning" is the warning from tip four, "every golf course deserves a long-range master plan" (Singer et al., 2009, p.26). When consulting with Audubon officials, a plan is required in the process. Further, "The Financial Equation" from tip five fits with the investment of using Audubon services and improvement suggestions as an investment opportunity for greater player enjoyment, and more efficient resource use. Strong economic incentives, where applicable for course profitability, can drive reduced course impact. The finally three tips all reinforce the usefulness of Audubon certification: the he certification influences "6. Delivering the Best Possible Product… 7. Management format…[and] 8. Developing New Customers" (Singer et al., 2009, p.27). Specifically the long term goal of ecological courses is attracting new customers and continuing to meet their standards year after year with an environmentally sensitive management plan.

Golf courses provide quality wildlife habitat, and stop urbanization or intensive agriculture (Leclerc et al. 2005). Nevertheless, are golf courses the best possible land use choice? Case examples of other uses from Tyler Grooms research include an open space preserve, organic farm, urban park, and community center. Municipal planners need to decide if a golf course is the right development decision. Audubon International consults with planners as well as managers in order to achieve conservation certification in which the courses are environmentally sensitive conservation resources and protect wildlife. This may influence the future of golfer preferences because reducing management lowers operating cost and, likely, the price to play.

Course Benefits

Course design can maximize functionality beyond resource conservation and wildlife protection. For example, Arlington lakesstorm and flood control course uses lakes specifically designed for flood control, that also contribute to the character of the course (Grooms, 97). The San Antonio use of a quarry exemplifies how courses can function to reclaim lands degraded from other uses (Grooms, 97). Pelican Lakes Golf course in Colorado replaced an old quarry in Windsor, and also has the advantages of rock out croppings, and natural rainfall collection pools (Grooms, 97). In addition, the ecological benefits of courses are greater than many other land use choices. The better conservation practices on courses limit the negative impact of course operation. Conservation practices can also lower operating cost through better resource use, as well as boost revenue with environmentally-minded consumers. Overall, golf courses are an ecologically functional business. "While golf courses often cost more than other amenities, their returns are consistently much higher" (Grooms, p.88). These returns also include nonmonetary ecological and social benefits. The recreation value for players and their learning experiences also indicates social benefits that are only partially represented in the market value.

Social Benefits

Players Value

Players value their time at least 45 dollars per 18-holes on the two courses under examination, according to their willingness to pay. Golf functions as a long standing form of recreation, competition, and exercise. The game increases mental focus, persistence, body awareness, and patience. Many players also enjoy the time outdoors socializing with friends or family. Wildlife experiences also add to the overall round of golf. Colorado players are possibly environmentally motivated when making their playing decisions which may influence players to choose an Audubon certificated course.

Societies Value

Although golf courses provide aesthetics and recreation to their surroundings, Seabrook golf course did not have a significant effect on housing price (Pompe et al.). However, golf has the potential to promote values in players that better themselves, and how they conduct themselves in everyday life. For instance, a person's dedication, practice, and progress will likely influence the achievement of goals beyond the game of golf. The activity is a natural stress reliever, and healthy light exercise activity. Environmental stewardship from replacing divots and preserving turf, as well as player behavior and etiquette, are desirable educational opportunities for adolescent golfers. Especially, when compared to other physical and more violent sports, golf offers an appropriate recreational alternative.

City Land Asset

The limited access to golf courses compliments conservation goals. Excludability counters the free rider problem and a more environmentally efficient outcome is expected when the price of play includes externalities in the market price. Preserving courses as a population sink enhances the effect of surrounding national parks, or county open space for breeding source groups. Golf course land offers yearly revenue for municipalities, in addition to the land holding with future value. Constituents enjoy golf, but non-players can benefit from other design strategies, like flood control or sound buffering. Cities can also mitigate excess storm runoffs impact on their infrastructure through using the waste water to hydrate the turf grass. Golf courses, when used correctly, add value and functionality to communities, and are an appropriate city land use.

Course Sustainability

Replacing Golf Carts

The 'Segway personal transporter' is currently in a trial at one Colorado golf course to replace golf cart use. Courses require this transportation to supply accessibility to elderly and handicap players. This transition eliminates a single rider in a two person cart, and likely the total energy used due to compact design and a direct line of travel to each individual's shots. Further, the overall quantity of materials in manufacturing are also less than traditional golf carts. These devices also can act as an attraction to new players who have interest in riding one of these personal transporters. Even less impactful playing styles are pull carts or bags that distribute the weight with backpack straps. These methods that involve walking also promote a healthier lifestyle.

Sustainable Design

Playing areas have the potential for a construction plan that maximizes the local ecology, and ecosystem services. Maximizing the usefulness of the course can justify the land use. Turf grass systems on compromised land benefit from vegetative reclamation. Soil resources are protected when correctly managed, and social or economic gains can offset other resource expenditures. These qualities are useful for Audubon certification, and encouraged for sustainability.

A course operating as a social and ecological system will reach a relative equilibrium with minor changes from design and management decisions that can shift the balance. Side-effects and negative interactions need preliminary trial areas, which entail monitoring. The effort is justified, because finding the appropriate resource requirements is essential to more environmentally sensitive golf courses. Other factors such as timing applications and species interactions are locally unique, but when properly incorporated into a course these improve the social and ecological functionality of the space. Audubon International's consultation process evaluates criteria like these when designing management plans. The more the properties of these attributes are understood the better the likelihood that certain characteristics of courses appropriately match design and management varieties.

Research and Education for Progressive Management

Collaborative assessment of healthy golf course practices, as well as adapting management, are required. The complex social and ecological systems framework fits appropriately with the complex interactions on a course. Further, locational variability means that no one solution is applicable on the

numerous global golf courses. Golf course managers need inter-discipline education in order to analyze the data and structure their plan to best use the locations natural features. Remote sensing and digital mapping give useful data to inform areas of improvement, and success in a management plan.

Locational Concerns

Specifically tropical climates and high temperature deserts should restrict golf course construction. The problem with tropical climates is the heavily leeched soils. The heavy rains strip nutrients, and the highly weathered soil shows low cation exchange capacity. A soils cation exchange capacity is the ability for the particles to hold on to positively charged ionic nutrients. Tropical soils are notoriously the worst, and require excessive fertilizer applications for both agriculture and turf grass systems. Further, course development tremendously lowers the biodiversity of these more ecologically diverse areas. The course acts as a matrix border to the forest that also changes the microclimate of the edge and limits species presence in the edge zone.

High temperature deserts are not the place for turf grass systems due to their higher levels of water consumption. The high heat and low humidity create substantial potential evapotranspiration, the actual evapotranspiration only consumes as much water is available. Golf courses import unnecessary quantities of water in these wasteful conditions. Golf course land use in deserts is not a sustainable resource decision. These courses do not benefit from minor improvements and are discouraged for these environmental reasons.

Policies to Ensure Societies Collective goals

Even the worst pesticides need not always be banned, but these pesticides require taxes to minimize their use. Environmentally detrimental chemicals do qualify for some high use values. However, a tax on these same chemical will discourage any excessive or unnecessary use. Requiring courses to modify their pesticide budget, or sacrifice quality, if prices are too high in order to increase demand and remain viable. Audubon International consults with golf courses to minimize the environmental impact of chemical applications, while maintaining turf playability. Their goal is to identify the best management practices for specific golf course localities. Golf course managers may also require congressional restriction of specific products to maintain environmental quality.

Some chemicals have worse impact than others, requiring research and monitoring to identify potential problems. The herbicides *2, 4-D, MCPP*, and *Fenarimol* have incomplete databases to categorize threat. Strong caution is advised when using the fungicides *Anilazine, Benomyl, Chlorothalonil,* and *Maneb* (Horsley et al., 1990). The two herbicides warranting strong caution are *Dicamba* and *Dacthal* (Horsley et al. 1990). These cautions relate to the chemicals' potential threat to ground water. Interestingly, no insecticides are listed as having an incomplete database, or being a threat. However, this study was limited to ground water threat, and insecticides may affect other species. Species threat creates another type of negative unintended consequence in addition to ground water contamination. These consequences and others need collaborative research in order to categorize the different levels of damage that chemicals may cause in order to apply the appropriate taxes. The taxes work to combat the market failure of environmental impacts not considered in the sales price. Also, the tax can change behavior such as when managers consider how to reduce operating cost. These regulations can also improve the health of the neighbors and users of golf course facilities.

Case Study

Course A and B: Similarities

Both of these courses are part of Westminster's city water reclamation program, which is recognized for environmental stewardship in the state of Colorado, and receiving accolades from the U.S. Environmental Protection Agency. These courses benefit from a well-designed municipality that integrates these facilities, and genuinely wants to preserve environmental resources. The main benefits of the water reclamation project complement golf courses and parks in the semi-arid climate. Human and wildlife health are not compromised with reclaimed water use, "In fact, reclaimed water quality standards are more stringent than those for surface streams, rivers, and irrigation channels" (City of Westminster Reclaimed Water). Turf is healthier from the nutrients contained in reclaimed water, and society benefits from a reduced demand on drinking water. This is a good model for water conservation in a municipality.

The two courses are roughly 6.5 miles apart, and both face the same competition. The multiple courses in the area act as substitutes and can force change at undesirable courses. So, the problem is identifying how golfers make their decision where to play. Course B is located near a larger course that is run from a parks and recreation district of the federal government. Course A faces a similarly large neighboring course, but this one is private and much more expensive. The short distances separating these courses do not seem significant to isolate players. If players make their decisions on environmental quality, than course A should attract more players.

Although individual managers differ from course to course, the superintendent and budget are from the same municipality. Major decisions regarding these courses come from the same place. None of the operating details of either course are a secret to the other. Meaning, similar course considerations, players, and price are all equal. The major differences are the design modifications and strategies to make Course A more ecologically situated in the landscape.

Course A and B: Differences

The Audubon Certification goals for Course A style signature courses are expressed as follows.

"The Audubon International Signature Program is an environmental education and assistance program created to help landowners and managers to follow sustainable resource management principles in a comprehensive manner when developing and then managing properties. The program aims to add and integrate wildlife conservation, habitat restoration and enhancement, water conservation and water quality protection and other areas of environmental protection and improvement with the other objectives for new developments. The Signature Programs' long term goal is to foster a stewardship ethic that leads landowners and managers, consultants, and the community to consider environmental, rather than just economic, costs and benefits in their decision making and to apply these environmental values routinely in land management" (City of Westminster).

The question is: are these sufficient for sustainability? If so, these courses need to address marketing to get the attention of players with environmental preferences.

http://www.ci.westminster.co.us/ParksRec/GolfWestminster/HeritageatWestmoor.aspx

Suburban development is prominent around course B as it was built within a housing community. Course A is adjacent to open space and has a naturalistic layout. This open space is a source of revenue, because Course A receives stipends from the Jefferson County Open Space Fund. This additional revenue source does not exist on Course B. The open feature of Course A is designed for ecological needs, but also contributes to economic security. This reflects a naturalistic course, which incorporates the surrounding environment.

Course A is a functioning sound and safety buffer between the County airport and housing subdivisions. The soft surfaces absorb sound, unlike some structural development. The area also hosts a yearly air show. The course is a preferable alternative to other developments in lieu of a crash, or other unfortunate accident. The economic vulnerability of a plane crash is much less on the course than for structural developments.

Further, Course A better utilizes trees on the course. Course B has less control of how the neighboring home owners plant trees and manage their lands. Course A has control of the area and Audubon and other recommendations are apparent with trees located on the front nine along a riparian basin. Trees accommodate some wildlife species, which are not overly abundant as to be a resource burden.

These design differences show the influence of environmental sensitivity, conservation certification, and multi-use planning of golf courses. The progress shown on the Audubon International signature golf course relates to the success of golf course certifications for attaining land use sustainability. Other managers can have confidence in the certification cost, because of the added viability it serves on other courses. The marketing of this conservation certification is encouraged to allow golfers to exhibit their environmental preferences. Promoting these golf courses as environmentally friendly is not just 'green washing' because the social and environmental benefits are associated with environmentally sensitive management.

Economic Comparison

The data shown below compares the two courses' revenue and expense for three fiscal year periods, 2008-2010. From left to right are 2010, 2009, 2008 with Course B revenue, then the same three years for Course B's expenses. The same two sets of information for Course A follows. The Course expenses are considerably similar. This information alone however does not discredit the Audubon Certifications economical resource incentives, but rather that the techniques the municipality learned for the signature course are now applied to all of their golf management. The ability to apply what is learned at the certified course to other course increases the municipality's benefit of pursuing certification. However, any non-cost effective environmentally beneficially practices likely are not

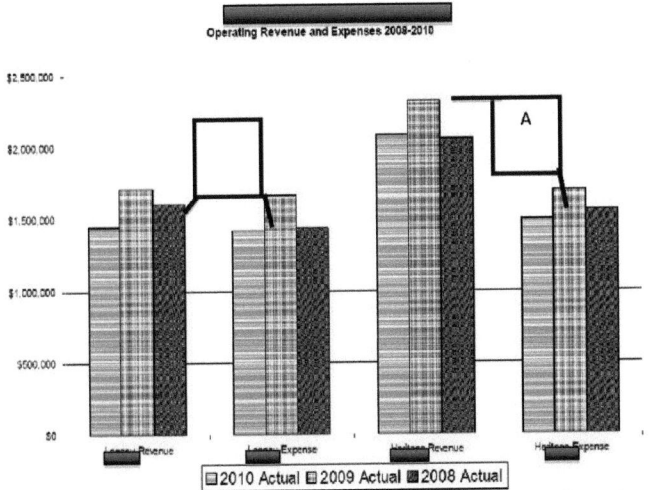

The variance between years is primarily due to the 2009 lease financing of a new golf cart fleet.

applied. The most significant difference is course revenue. An investigation into consumer preferences could determine if environmentally friendly golf courses are currently more desirable. The trend of environmentally sensitive golf courses will likely only continue when the consumers reinforce the trend. The future of ecological courses is mostly influenced by demand for naturalistic golf courses, as well as consumer confidence in the certification agencies. The most important missing information on potential sustainability is the attitude of the consumers.

Recommendations

What does this mean for Colorado golf courses?

Ecologically healthier golf courses for consumers and communities may result from certification. Colorado's climate does not facilitate traditional golf courses. Designs with high tree density and lush sprawling turf grass are suitable in moist coastal climates, however, the high altitude, semi-arid moisture regime, and numerous cloudless sun-filled days stress turf areas in Colorado. Using reclaimed water is specifically appropriate in Colorado, but everywhere can benefit from the reduced demand on purified water. Conservation consultants, like Audubon International, specialize in ecological preservation and design that is site-specific for management plans. These strategies utilize chemical management and water schedules that compliment healthy turf, and also reduce impact on wildlife. The practice of efficient resource use encouraged in conservation certification reduces operation cost, in addition to minimizing externalities to wildlife and other ecosystems.

Certification consultants recommend best management practice resource use for individual courses to operate more efficiently and potentially even sustainably. Goals include identifying better solutions to common turf problems rather than superficial counter measures. Wetting agents are commonly used, but microbial fungus is the most likely cause of hydrophobic soils. Conservation consultants can identify appropriate fungicides for Colorado, and their experience can benefit all the courses they work with in similar circumstances. The knowledge of successful ecological features from previous certified courses helps ensure the Audubon process for the future. When they work from past success, and learn from failures, Audubon International can more fittingly advise golf courses each successive year.

Conclusions

Certification, in combination with environmental consumers, makes a course more competitive and viable. However, consumers need both awareness of ecological courses, and personal motivation to choose environmentally responsible courses. If demand does not respond to the supply of environmental courses, then the trend of paying for conservation intervention and certification will not continue. Preliminary understanding of consumers' reflect imprudent response to environmental concerns.

Growth regulators are a better choice than wetting agents for chemically increasing water use efficiency for turf grass systems. Growth regulators chemically instruct the plant to focus new growth to the roots. With higher root activity, soil becomes more permeable for water and the root mass more adequately utilizes soil resources. This also allows grass to quickly heal from surface trauma. The other result is less leaf growth, which reduces both evapotranspiration and trimming. This technique conserves water as well as fuels resources used for course maintenance. Responding to hydrophobic soils does not require wetting agents. Healthy microbial communities and/or fungal mitigation can prevent the water repellency entirely (Hallet et al., 2001). Modern golf course management needs to shift to identify methods for using environmentally sensitive strategies that both minimize externalities and conserve resources.

Course design requires digital mapping to appropriately situate playing areas within a specific location. Consequently wildlife is better accommodated with planned corridors. Also, sensitive habitat is better preserved. Buffers and irrigation systems operate more adequately when placement is planned to fit the landscape. Advances in remote sensing give courses a technological advantage that, when applied properly, will assist planners in protecting the course resources and surrounding habitats. Old courses can benefit from consultation on these modern techniques and consider redevelopment with environmental priorities as well as economic goals. Social and environmental benefits currently are not considered in the market value, yet add to the value of golf course land uses. Communities can benefit from living near courses without actually being golfers.

Reclaimed water is an additional strategy for using course attributes and resource requirements more collaboratively with the community. The city's water systems demand is reduced with the diversion of waste water to valuable non-potable uses. The courses benefit from remaining nutrients not stripped through a purification processes. The only side effect of reclaimed water is salinity build up in the soil. When properly managed, the salinity is positively useful in a turf system as a weed suppressant.

Pesticides require some type of government regulation related to distribution in order to minimize widespread negative effects. The most environmentally damaging products should not be used indiscriminately. These products could require heavy taxing in order to restrict their purchase to their highest use values. Total bans on these substances may cause a problem with an inadequate substitution for a specific use. Taxes influencing the price may deter purchasing and lead managers to consider purchasing less expensive and environmentally friendly products.

Society benefits from the recreational value of golf, and the values taught in golf. Golf teaches the importance of practice for self -improvement. Additionally, the game encourages etiquette, integrity, and environmental stewardship. Further, golf is a good form of outdoor recreation, and is a healthy physical activity. Also, many sporting alternatives are more physically demanding and aggressive. Therefore, players can benefit society from the principles and values educated in golf.

To further progress golf synergies, will require more collaborative interdisciplinary investigations. Audubon International's success depends on their effort to involve environmental scientists, city planners, golf course developers, professional players, and other stakeholders. The amalgamation provides an encompassing perspective to execute an ecologically certified course. Tree and turf specialist have an important role in identifying appropriate management techniques with reduced environmental impacts.

Golf needs only a few improvements for sustainability, because the game already has substantial economic, societal, and environmental value. The steps golf course managers need to make for sustainability are resource concerns. First, water is essential to all life, and golf needs to preserve this vital resource. Some courses are water wasters and/or contaminators; the certification procedure does not allow this. Second, are golf's energy requirements. Each course requires mechanical maintenance that consumes fuel resources. Carts and clubs both require high energy construction, and the carts need charging on a daily basis. Audubon International focuses on managing these resources, and conserving the resources, in turn, can reduce the course budget.

Audubon International certification is a necessary start for promoting sustainable golf course management. The agency adds institutional diversity for environmental accountability for golf courses. Ecological sensitivity was previously lacking in golf course development. Audubon International is necessary to change this because they save courses money through best management practices that conserve resources. Monitoring resources actually increases the courses profitability. The potential effect on how conservation certification influences consumers is the primary concern for investigation. The success of maintaining conservation in golf is how consumers respond. With the awareness of environmental protections provided in a conservation certification, consumers need confidence in the agency. This investigation finds strong support for Audubon International's ecological contributions to participating courses. The other aspect of concern is how to motivate people to make more ecological decisions about where they will pay to play.

Bibliography

Audubon International. *Golf's Green Bottom Line: Uncovering the Hidden Business Value of Environmental Stewardship on Golf Courses.* A Research Project of Audubon International. 2007.

Audubon International. *Proceedings of the Sustainable Communities Summit, 2008.*

Audubon International. *Lessons for Eco-Design & Development, 2009.*

Aplet, J. & Conn, D. (n.d.). The uses of completed landfills. *Planner for the Southern California Association of Governments,* 1-10.

Andersson, K., Burns, M., Bursztyn, M., Henry, A. D., Laudati, A., McNie, M., & McNie, E. (2008). What is sustainability science?. *The Ruffolo Curriculum On Sustainability Science: 2008 Edition,* 1-3.

Andersson, K., Burns, M., Bursztyn, M., Henry, A. D., Laudati, A., McNie, M., & McNie, E. (2008). Values and principles in sustainability science. *The Ruffolo Curriculum On Sustainability Science: 2008 Edition,* 1-4.

Arya, L. , Bowman, D. , Thapa, B. & Cassel, K. (2006). Soil science society of America Journal. *Soil Science Society of America, 72*(1), 25-32. DOI: 10.2136/sssaj2006.0232

Baird, J. H. (1998, September). Reducing pesticide and nutrient runoff using buffers. *Golf Course Management,* 57-61.

Beard, J. (1996, March). The benefits of golf course turf. *Golf Course Management,* 57-61.

Carrow, R. N. (2006). Can we maintain turf to customers' satisfaction with less water?. *Agriculture Water Management, 80,* 117-131.

Carrow, R. N. , Duncan, R. R. & Wienecke, D. (2005, July). Bmps approach to water conservation on golf courses. *Golf Course Management,* 73-76.

Chawla, L., & Williams, S. S. (2011). *Child friendly lawns and gardens.* Informally Published manuscript, Children, Youth and Environments Center for Community Engagement, University of Colorado, and Denver, U.S.A.

Cisar, J. University of Ft. Lauderdale Research and Education center. (2004). New directions for a diverse planet. *Proceedings of the 4th international crop science congress.* CDROM:

Clark, W. C. (2007). Sustainability science: A room of its own. *PNAS, 104*(6), 1737-1738.

Colding, J., Lundberg, J., Lundberg, S. and Andersson, E. (2009) Golf courses and wetland fauna. Ecological applications. 19(6), pp. 1481-1491.

Couillard, A. & Wiecko, G. (1998, May). A saline solution: Seawater as a selective herbicide. *Golf Course Management*, 54-57.

Cox, C. (1991). Pesticides on golf courses: Mixing toxins with play. *Journal of Pesticide Reform, 11*(3), 2-4.

Fry, J. & Hongfei, J. (1998, November). Plant growth regulators may help reduce water use. *Golf Course Management*, 58-61.

Gange, A. , Lindsey, D. & Ellis, L. (2001). Can arbuscular mycorrhizal fungi be used to control the undesirable grass poa annua on golf courses?*Journal of Applied Ecology, 36*(6), 909-919. Retrieved from http://onlinelibrary.wiley.com/doi/10.1046/j.1365-2664.1999.00456.x/full

Gange, A. , Lindsay, D. & Schofield, M. (2003). The ecology of golf courses. *Biologist, 50*(2), 63-68.

Garling, D. & Boehm, M. (2000). Temporal effects of compost and fertilizer applications on nitrogen fertility of golf course turfgrass. *Agronomy Journal, 93*(3), Retrieved from https://www.agronomy.org/publications/aj/articles/93/3/548

Grooms, T. (n.d.). Alternatives to golf course developments in an environmentally sensative market. *Cornell Real Estate Journal*, 88-99.

Hallett, P. , Ritz, K. & Wheatley, R. (2001). Microbial derived water repellency in golf course soil. *International Turfgrass Society Research Journal, 9*, 518-524.

Hodgkison, S. , Hero, J. & Warnken, J. (2007). The efficacy of small-scale conservation efforts, as assessed on Australian golf courses. *Biological Conservation, 136*, 576-586.

Hueber, D. (2010). "Code blue" for U.S. golf course real estate development: "Code green" for sustainable golf course redevelopment. *Journal of Sustainable Real Estate*, 1-42.

Karnok, K. J. (2006, July). Which wetting agent is best?. *Golf Course Magazine*, 82-83.

Karnok, K. J. & Beall, M. (1995, August). Localized dry spots caused by hydrophobic soils: What have we learned. *Golf Course Management*, 57-59.

Karnok, K. J. & Tucker, K. (2002, July). Water-repellent soils part 1: Where are we now? *Golf Course Management*, 59-62.

Karnok, K. J. & Tucker, K. (2002, July). Water-repellent soils part 2: More questions and answers. *Golf Course Management*.

Karnok, K. J. & Tucker, K. A. (2003, June). Turf grass stress, water-repellent soils, and LDS. *Golf Course Management,*

Karnok, K. J. & Tucker, K. A. (2008, June). Using wetting agents to improve irrigation efficiency. *Golf Course Management*, 109-111.

Karnok, K. J. , Xia, K. & Tucker, K. A. (2004, June). Wetting agents: What are they and how do they work?. *Golf Course Management*, 84-86.

Kates, R. W., Parris, T. M., & Leiserowitz, A. A. (2005). What is sustainable development? Goals, indicators, values, and practice. *Environment, 47*(3), 9-21.

King, K. & Balogh, J. (1999). Modeling evaluation of alternative management practices and reclaimed water for turfgrass systems. *Journal of Environmental Quality, 28*, 187-193.

King, K. , Balogh, J. , Hughes, K. & Harmel, R. (2007). Nutrient load generated by storm event runoff from golf course watershed. *Journal of Environmental Quality, 36*, 1021-1030.

King, K. , Harmel, D. , Torbert, A. & Balogh, J. (2001). Impact of a turf grass system on nutrient loadings to surface water. *Journal of the American Water Resource Association, 37*(3), 629-640.

King, K. , Hughes, K. , Balogh, J. , Fausey, N. & Harmel, R. (2006). Nitrate-nitrogen and dissolved reactive phosphorus in subsurface drainage from managed turfgrass. *Journal of Soil and Water Conservation, 61*(1), 1-10.

Larsen, S. & Fischer, J. (2005). Turfgrass management and weed control on golf course fairways without pesticides. *International Turfgrass Society Research Journal, 10*, 1213-1221.

Leclerc, J. , Che, J. , Swaddle, J. & Cristol, D. (2005). Reproductive success and development stability of eastern bluebirds on golf courses: Evidence that golf courses can be productive. *Wildlife Society Bulletin, 33*(2), 483-493.

Lewitus, A., Schmidt, L., et al. (2003). Harmful algal blooms in South Carolina residential and golf course ponds . *Population and Environment, 24*(5), 387-413. DOI: 10.1023/A:1023642908116. http://www.springerlink.com/content/h01607643267j1n3/

Ludwig, F. , Dawson, T. , Prins, H. , Berendse, F. & Kroon, H. (2004). Below-ground competition between trees and grasses may overwhelm the facilitative effects of hydraulic lift. *Ecology Letters, 7*, 623-631.

Mallin, M. , Ensign, S. , Wheeler, T. & Mayes, D. (2000). Pollutant removal efficacy of three wet detention ponds. *Journal of Environmental Quality, 31*(2), 654-660. DOI: 10.2134/jeq2002.6540

Mankin, K. (2000). An integrated approach for modeling and managing golf course water quality and ecosystem diversity. *Ecological Modeling, 133*(3), 259-267. DOI: 10.1016/S0304-3800(00)00333-1

Miranda, J. (2005). *Sustainable reclamation in golf course design.* Informally published manuscript, Department of Landscape Architecture, Ball State University, Muncie, IN.

Miles, C. & Running, S. (2005). Mapping and modeling the biogeochemical cycling of turf grasses in the United States. *Environnemental Management, 36*(3), 426-438.

Miller, C. (2001, April). Comparing wetting agents: Llong-term vs. short-term. *Golf Course Management,* 60-64.

Mintenko, A. & Smith, R. (1999, November). Evaluation of native grasses for low-maintenance turf. *Golf Course Management,* 60-63.

Nelson, M. (1997). Establishing natural areas on the golf course. *USGA Green Section Record, 35*(6), 7-11.

Oostindie, K. , Dekker, L. W. , Wesseling, J. G. & Ritsema, C. J. (2008). Soil surfactant stops water repellency and preferential flow paths. *Soil Use and Management, 24,* 409-415.

Pompe, J. J., & Rinehart, J. R. (n.d.). The effect of golf course location on housing value. *The Coastal Business Journal, 1*(1), 1-12.

Qian, Y. & Mecham, B. (2005). Long-term effects of recycled wastewater irrigation on soil chemical properties on golf course fairways. *Agronomy Journal, 97,* 717-721.

Sifers, S. L. & Beard, J. B. (1999, September). Drought resistance in warm season grasses. *Golf Course Management,* 67-70.

Singer, R. & Richardson, F. (2009, January). 8 tips for improving municipal golf courses. *Parks & Recreation,* 23-28.

Smiley, R. (1981). Non-target effects of pesticides on turf grasses. *Plant Disease, 65,* 17-23.

Tanner, R. & Gange, A. (2005). Effects of golf courses on local biodiversity. *Landscape and Urban Planning, 71,* 137-146.

Terman, M. (1997). Natural links: Naturalistic golf courses as wildlife habitat. *Landscape and Urban Planning, 38,* 183-197.

Westminster City Council Agenda, Westminster Economic Development Authority. (2011). *8.a. consent agenda financial report for December 2010* Retrieved from http://www.ci.westminster.co.us/agendas/default_2275.htm

Wheeler, K. & Nauright, J. (2006). A global perspective on the environmental impact of golf. *Sport in Society, 9*(3), 427-443.

Whitney, M. (2011, January). Golf courses more than a walk in the park. *Parks & Recreation*, 75-77.

Winter, J. G. & Dillion, P. J. (2005). Effects of golf course construction and operation on water chemistry of headwater streams on the Precambrian shield. *Environmental Pollution, 133,* 243-253.

Winter, J. G. , Somers, K. , Dillion, P. , Peterson, C. & Reid, R. (2001). Impacts of golf courses on macroinvertebrate community structure in Precambrian shield streams. *Journal of Environmental Quality, 31*(6), 2015-2025. DOI: 10.2134/jeq2002.2015

Yao, H. , Bowman, D. & Shi, W. (2006). Soil microbial community structure and diversity in a turfgrass chronosequence: land-use change versus turfgrass management . *Applied Soil Ecology, 32*(3), 209-218. DOI: 10.1016/j.apsoil.2006.01.009